Making Perfume By The Quart

A do-it-yourself project book

Illustrated with 28 professional color photographs

Philip Goutell

Lightyears, Inc.
Walden, NY 12586
www.PerfumeProjects.com

Contents

Illustrations

Photography by Philip Goutell

Author's Note

This book is a step-by-step guide to making one U.S. quart of perfume in the United States. The materials used in this do-it-yourself project have all been obtained either from sources in the United States or from international sources that ship to the United States. The alcohol used was denatured by a U.S. Treasury Department standard for use in perfume. The labeling instructions found in Chapter 9 are from U.S. Food and Drug Administration regulations. Readers outside the United States are encouraged to look into their own country's regulations and available sources of required materials.

Philip Goutell
July/August 2021

Introduction

If you seriously want to create a perfume developed by you, named by you, and offered for sale by you, you can do it — if you follow the steps described in this book. You won't need much money and you can get started any time you're ready. You can get started right now.

One quart only

The following chapters are a step-by-step guide to guide you through the process of making one quart of your own perfume. The "instructions" are not hard to follow. If you live in the United States you should have no trouble obtaining the supplies you will need, and they won't cost all that much.

But, to create your quart of perfume successfully, you must avoid getting sidetracked. You must avoid trying to rewrite this book (which would be easy to do!) Keep your eyes open and take each step in order. No side tripping. Stay focused and you will come out with a quart of your own perfume, bottled, capped with fine mist spray pumps, correctly labeled, and ready to sell.

Why just one quart?

If your project were to involve a major investment it is likely you would add an experienced guide to your team to walk you through each step of the perfume development and manufacturing processes. If all were to go well — and sometimes it doesn't — you would come out with a number of cases of a very professional looking product, nicely bottled, nicely packaged, and with a nice aroma. If your guide was good you would also come out with a perfume produced at a minimal cost per bottle, giving you the opportunity to achieve a high markup. Yet all this comes with no guarantee that your perfume will sell. Launching a new perfume always involves risk.

Making a single quart of perfume takes you through all the steps involved in creating a new perfume, whether the investment is peanuts or $20,000 or more. If you are simply curious about *how* a perfume goes together, this is your opportunity to learn. If you have already made your own perfume oil and want to see it developed into a bottled, ready to sell product, this is your chance to do it while spending perhaps less than what you already spent developing your oil.

If you are, or would like to become, a professional marketer, this project gives you the opportunity to learn perfume production while spending just pennies and, for this small investment, come out with a bottled, ready to sell fragrance that can be used for test marketing or presentations to investors.

You can start organizing your project today!

The most powerful reason for setting out quite deliberately to make just one *quart* of perfume is that it removes all the obstacles that might be holding you back from making *any* perfume. To make just one quart of perfume you don't need to find investors, you needn't worry about losing money, you don't have to take your idea to a committee to have it vetted. You don't have to wait for anyone's decision. You can get started right now. Nothing is holding you back.

Chapter 1
Here I describe everything you need to do

Look at this project working backward, from finish to start. You will come out with a number of bottles filled with your own perfume dispensed by spray pumps mounted on the bottles. We'll leave off the box for now as that requires some discussion. You will find that discussion in Chapter 10. The spray pump — a real requirement for this project — will be covered in Chapter 7.

The goal of this project is for you to produce from 16 to 64 bottles of your own perfume that will look something like these examples.

Now, how did we get to that finished perfume? We filled each bottle using a turkey baster with a metal tip with perfume we had poured into a 4-cup glass measuring cup from a quart bottle in which a mixture of fragrance oil, water, and alcohol had been aging for a month. You can see already that it's pretty simple.

The fragrance oil, water, and alcohol that went into the quart bottle was carefully measured out using the same 4-cup measuring cup we used when filling the individual bottles. The entire project requires only these supplies and materials:

Metal tipped turkey baster used for filling perfume bottles. The metal tip is very much like the tip used in both automated and manual filling machines. This instrument is particularly useful for filling sprinkler neck bottles which are difficult to fill with any other device.

Quart bottles, three or four of them

Fragrance oil (the perfume... the scent!)

Alcohol

Water

4-cup glass measuring cup

Funnel

Metal turkey baster

Perfume bottles

Spray pumps

Filter paper

Funnel used to transfer liquids from measuring cup to quart bottle. Either metal (like this one) or glass will be fine. Avoid plastic.

That's it. No expensive equipment needed. You can do it on a desktop or kitchen table. Your two biggest obstacles will be obtaining the fragrance oil you want for this project and obtaining the alcohol you will need. These obstacles don't amount to much. Getting your fragrance oil is covered in Chapter 2 and getting your alcohol is covered in Chapter 3.

Reverse side of measuring cup showing graduations in milliliters (left) and deciliters (right).

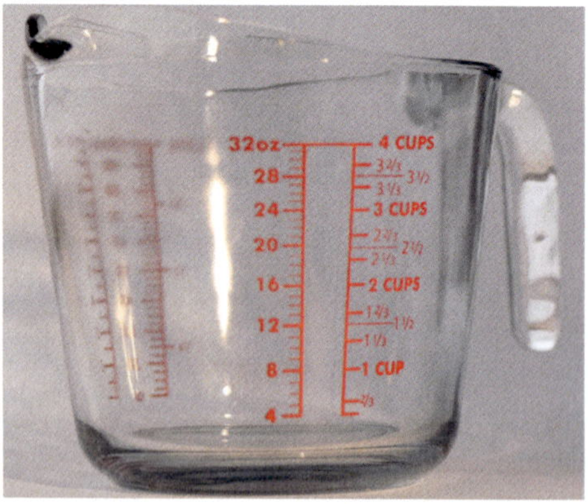

4-cup glass measuring cup. This side is graduated in fluid ounces and cups.

So the next two chapters will get you over the biggest hurdles to making your own perfume. After that it's all pretty mechanical and routine. Our sources for the supplies and materials used in this project are found in Appendix A.

Chapter 2
Getting your fragrance oil

The heart of your perfume is your fragrance oil, the scent itself. If you have made a perfume oil of your own and you like it, use it for this project. If you have a friend who has made a perfume oil you like, perhaps your friend will give you some for this project. If you don't have a perfume oil that you or a friend have made you will have to buy your oil.

If all you want is enough oil for this project, there are many ready made oils available online. They are generally well made, not expensive, and will work well for this project. If, on the other hand, this project is a warmup exercise for a well funded marketing effort, you may consider, right from the beginning, working with a professional perfumer who can create an original scent for you.

Perfume oil you made yourself

You may be reading this book because you have created your own perfume oils and want to see how one of them translates into a finished, bottled product. This is a perfect opportunity for you. If you have been experimenting with essential oils and other aroma materials, this is a chance to find out how your creations stand up when given a professional treatment. If you haven't yet tried to make your own perfume oil but would like to, there are many websites with do-it-yourself instructions for making a simple perfume. The only warning I have is be sure to use one of the alcohols recommended in the next chapter. If you do this you shouldn't shouldn't have and problems with your perfume clouding up, no matter what essential oils and other materials you have used.

If you are a student of perfumery working with a range of professional materials — perhaps both synthetics and naturals — this is your opportunity to find out how one of your creations "works" when given a rest and then bottled. You may discover you have a real knack for creating fragrances and what you have created could be sold very nicely.

Using a fragrance oil made by a friend

If you haven't made a perfume oil you like on your own, you might have a friend who has made an oil you like. You might find a new friend with a good oil through social media. Your old or new friend might be willing to provide the oil you need for this project.

Before you agree to use someone else's fragrance oil to make your perfume, be sure you like it well enough to invest your time and money in it. Then be clear on the terms on which this oil is being supplied to you. Will you be able to call the finished product *your* perfume or will the person who supplied it to you want some or all of the credit? Will you be paying for it or will your friend or other creator be supplying it to you free? If your finished product proves a success, *will you be able to get more of the same oil?* The *exact* same? Here you'll want to know whether this fragrance oil was just a lucky, random mix or whether the creator carefully recorded the formula so that additional batches could be made. And, if this person recorded the exact formula, did they also record the source and perhaps even the batch number of each item used in their formula? You don't want the disappointment of making a second batch of your successful fragrance only to find, after you've produced another quart or more, that the oil you've been given the second time around *is not the same* oil that gave you success and now your fragrance is a colossal failure.

Unless you have a good sense for spotting good fragrances and have the good people skills needed to deal with your oil creator friend (where creativity and money are involved, you can get some very unexpected surprises!), this might not be the best way for you to obtain your perfume oil. On the other hand, if you are feeling adventurous...

Buying your perfume oil

If you don't have a perfume oil you've made yourself or have been given by a friend, you will have to buy your oil. There are two ways to do this. The first is quite practical for this project and quite inexpensive. You simply go online to one of the many websites that sell fragrance oils and purchase what you will need for this project. The second way is considerably more expensive but it is what you would do if you were making a major investment to launch a new perfume: you commission a professional perfumer to create an original fragrance for you.

Buying ready made fragrance oil

The simplest, least expensive way to come out with a good perfume of your own is to purchase a small quantity of a ready mixed fragrance oil. There are many sources on the internet. The price will be modest regardless of the oil you choose.

A wide variety of perfume oils are easily available online. Some are original compositions; most are the vendor's versions of popular designer fragrances. All will be made to professional standards and will be easy to work with. Once finished, you will give your fragrance a name of your own choosing and that fragrance then, essentially, becomes your own. Just be aware that when you buy one of these stock fragrances for your project, the same *oil* will be sold to others. But once you have *named and labeled* your perfume — in effect given it *your* own trade name — it will be very unlikely any competitor will make the connection between your perfume and the oil's source. Likely your perfume will be taken for an original.

The real issues are scent and continuity. Continuity refers to being able to get more of the same if you are able to sell your perfume successfully. This means you will want to find an established vendor who will have a continuous supply of the oil you select. You do not want to find yourself with a big winner and then discover you cannot get any more of the oil that is doing so well for you.

Selecting a scent will be your biggest challenge. There are dozens of fragrance oils on the market, many of which are similar to all-time favorites and currently trending perfumes. Here you have to think a bit about your goal. Are you simply making a perfume to please yourself or are you making a perfume you hope to share with others, either as a gift or for sale. If so, you must consider *their* tastes. What scents will *they* be interested in trying? If you feel a bit stumped at finding the answer here, you're not alone. Every major fragrance marketer holds their breath each time they introduce a new fragrance. They have thousands of dollars invested and, until the public reacts, they don't know whether they've created a success or a failure.

Commissioning an original fragrance from a professional perfumer

If you are looking at perfume as a business opportunity and your quart project is being undertaken to get a feel for how it's done, and to get a product to use for some test marketing, the only source you will go to for your oil will be an established F&F (Flavor and Fragrance) house that will introduce you to a professional perfumer who will create an original fragrance oil for you.

Here cost may not be the issue depending on the arrangements you can make. The vendor, your F&F house, makes their money by supplying perfume by the pound. They might offer to develop an original fragrance for you in return for your commitment to purchase a certain minimum number of pounds of the finished oil. The price per pound will either be based on the ingredients used (the generous use of naturals can make your oil quite expensive!) or an agreement on a maximum price per pound, leaving the perfumer to work up a formula that will meet that per pound guideline.

Keep in mind that unless you have made another arrangement in advance, the perfumer or F&F company will *own* the formula and be your only source for this oil, even though you were the one who gave them your input and paid to have it created. Your protection in this arrangement is that your perfumer or F&F company will not share this formula with others or make this particular oil for anyone else.

Scent seeking suggestion

If you plan to buy your oil but aren't sure what scent you want to go with, of course you can and should sample popular fragrances at perfume counters but also, by going to websites that sell fragrance oils, you can collect a bunch of samples and use them to test how you, your friends, and others react to them. This small investment can provide you with some good research data.

Fragrance oils, "home made" and commercial off the internet.

Chapter 3
Alcohol

NOTE: Laws and customs regarding alcohol vary from country to country and can even vary from region to region within countries. The following is focused on alcohol's availability and its regulation in the United States.

When you make your quart of perfume, about 90 percent will be alcohol and about 10 percent of the alcohol will be water. Right from the beginning, think alcohol. Forget about using a non-alcoholic solvent. Forget about using vodka or Everclear. For this project use one of the two specially denatured alcohols you'll read about below.

Specially denatured alcohol (SDA)

Denatured alcohol is alcohol which has been treated with "denaturants" that make it unfit to drink. This gives it tax and regulatory advantages. Commercial perfumes, whether made by mass marketers or indie perfumers, all use specially denatured alcohol (SDA).

The U.S. Treasury Department publishes formulas by which alcohol can be denatured. Of these denaturing formulas only a handful are authorized for use in perfume and only two of these formulas are suitable for this project.

SDA 39-C and SDA 40-B

Only two types of denastured alcohol are appropriate for this project and, in general, for making most types of perfume. These are designated "SDA 39-C" and "SDA 40-B". SDA 40-B is currently the popular choice but either will do. SDA 39-C requires that one gallon of diethyl phthalate be added to every 100 gallons of alcohol. SDA 40-B requires that one-sixteenth avoirdupois ounce of denatonium benzoate, N.F., and 1/8 gallon of tert-butyl alcohol be added to each 100 gallons of alcohol. Both of these formulas yield alcohol that is very nearly odor free.

When you shop for your alcohol in the U.S. you will be looking for one of these two denatured alcohols. Currently SDA 40-B is relatively easily available. Vendors who sell it will enlighten you on regulations and restrictions regarding its purchase and use but for purposes of this project the only restriction you will find is that alcohol is flammable and thus incurs some extra shipping charges.

Alcohols to avoid

Advocates of home perfumery have long suggested vodka or Everclear, a highly alcoholic beverage, as stand-ins for a correct professional alcohol. While it is understandable that a hobbiest might use one of these, as vodka in particular is easy to obtain, and they will work for a perfume not intended for sale, but if you're serious about the perfume you're making for this project you'll go through the trouble of getting the correct alcohol.

Pure grain alcohol

From specialty sources you may be able to purchase 200 proof, pure grain alcohol that has *not* been denatured. This will work just fine for *making* your perfume but it presents a problem if you plan to *sell* it. In addition to creating your perfume you will have created an alcoholic beverage. To sell an alcoholic beverage you need a liquor license. You *don't* want to go down this path.

Denatured alcohols other than SDA 39-C and SDA 40-B

All alcohol that is denatured is denatured to reduce taxation but not all alcohols that are denatured are denatured for the same use. Alcohol is denatured by various formulas for various applications. Denatured alcohol from one formula can be used as fuel for marine stoves. From another formula it can be a cleaning agent. From one formula it can be an aid in sanding by raising the grain. From one formula it can be added to windshield washer fluid. From another formula it can be used by exterminators. None of the alcohols denatured by these formulas can be used in a beverage, nor can they be used to make perfume. When you shopping for denatured alcohol be sure it is labeled SDA 39-C or SDA 40-B.

Rubbing alcohol

Rubbing alcohol comes in a number of varieties with various percentages of isopropyl alcohol or ethanol based liquids. You would not want to use rubbing alcohol for any perfume you intend to share with others. You will not use rubbing alcohol for this project. Rubbing alcohol is harsh on the skin and can give your perfume a bit of that "rubbing alcohol" smell.

Methanol (methyl alcohol)

Methanol should never be used in perfume. It is dangerous in a number of ways. On the skin it has a toxic effect. Drinking methanol can cause blindness or death.

"Perfumer's Alcohol"

The name "perfumer's alcohol" has been attached to many blends. It can be SDA 39-C or SDA 40-B but it can also be a mixture of ethanol and denaturants such as isopropyl myristate and monopropylene glycol. For this project these should be avoided. Be sure to read the label on any alcohol you purchase.

Water added

It is generally desirable to have some amount of water in your perfume and the water is generally added as part of the alcohol. Commercial filling houses might stock an alcohol of around 170 proof. This means that water has been mixed with 200 proof alchohol so that the mixture is now 85 percent alcohol and 15 percent water.

If you are able to obtain alcohol that has already been mixed with water so that it is now 170, 180, or 190 proof, you're fine. But if you are purchasing 200 proof alcohol you will to want to add water to it before you mix it with your fragrance oil.

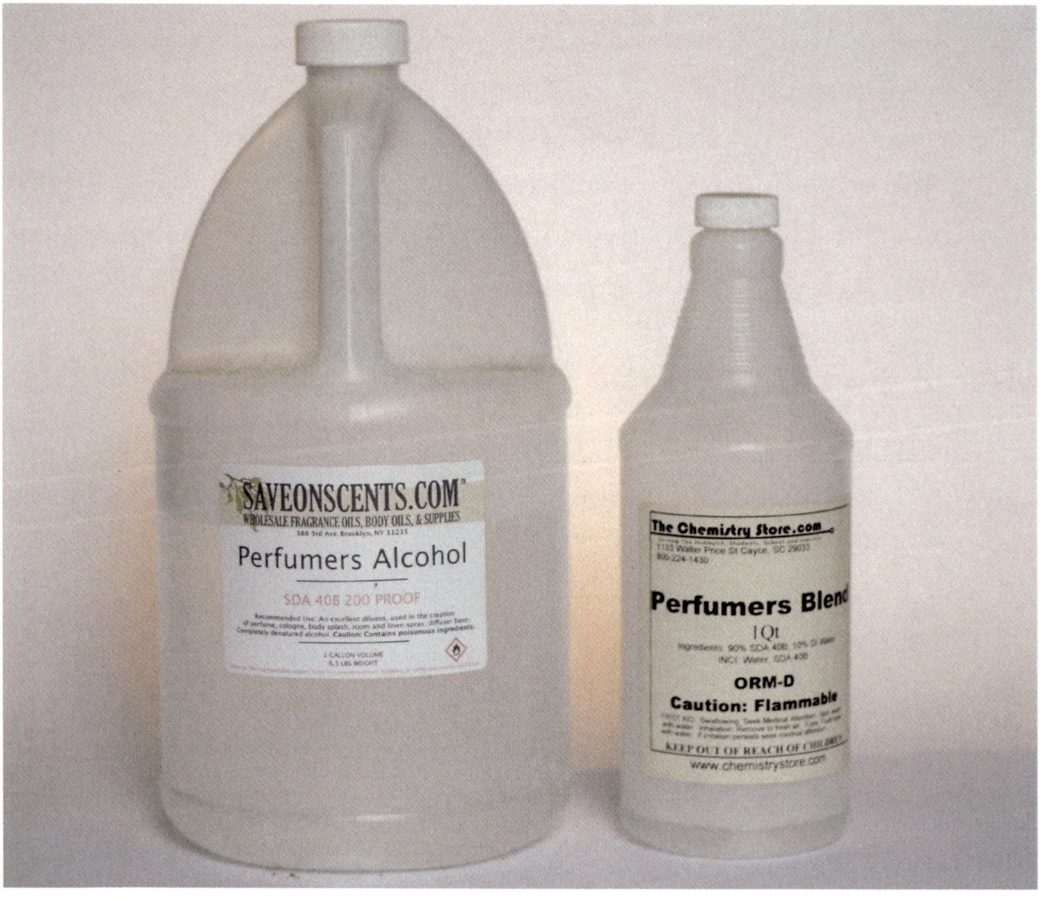

Alcohol SDA-40B. The jug on the left is 100% alcohol. The quart bottle at the right is 90% alcohol and 10% water.

Chapter 4
Water

If you are purchasing SDA 39-C or SDA 40-B that is less than 200 proof, water has already been blended with your alcohol. You don't need to add water. You don't need to buy water. But if your alcohol is 200 proof — pure grain alcohol (denatured) — you will want to cut it with water. Why? A very good question but no definitive answer.

Why use alcohol that has been mixed with water?

The only point agreed upon by most in the industry is that water in the alcohol improves the perfume. This may be because water increases the staying power of the perfume, particularly the top notes. It may be because water lessens the drying effect of the alcohol and makes the perfume gentler on the skin. Do your own experiments and decide for yourself. Going forward with this project we will use alcohol with water. It's as simple as that.

What kind of water to use?

Spring water is out. Too many minerals and other substances and not necessarily consistent from one batch to the next. The impurities in spring water could react with the materials used in your perfume oil. This is something you want to avoid.

Distilled water or deionized water?

You don't want elements in your water that will react with your perfume oil. Both distilled water and deionized water are purified waters. Neither are totally pure. With deionized water, ions that could be troublesome by reacting with your oil have been removed. With distilled water troublesome impurities have been removed by distillation. In double distilled water the water is distilled a second time, making it even purer.

Either distilled or deionized water are fine for this project. My personal choice has been deionized water because I know it is used in commercial perfumes.

Distilled water can be purchased at your local supermarket although you may have to look a bit to find it. Steam irons that once required distilled water now survive on tap water. Car batteries no longer require topping up with distilled water. A gallon of distilled water will cost you less than a quart of deionized water and no shipping charges are involved.

Deionized water is not particularly expensive but you will have to obtain it through a chemical or laboratory supply house and you'll likely have to purchase at least a quart which will be far more than you'll need for this project. Moreover, the shipping charge may be almost as much as the cost of the water.

How much water?

You'll want your alcohol to be at least 5 percent water (190 proof) and not more than 15 percent water (170 proof). For this project if you have to add water to your alcohol I suggest you settle on 180 proof — 10 percent water and 90 percent denatured alcohol.

Mixing alcohol and water

To mix your alcohol and water you'll need:

Quart bottle
Funnel
4-cup measuring cup

Fill your measuring cup with alcohol to the 29.80 ounce mark. Add water until it reaches the 32-ounce (1-quart) mark. Using your funnel, transfer your mixture from your measuring cup to your quart bottle. Shake it a bit and let it sit for a week. You now have one quart of 190 proof alcohol.

> NOTE: While you're at it you might want to make more than a quart of 190 proof denatured alcohol. Gallon jugs are generally sold in cases of four. If you decide to purchase a gallon jug and want to fill it with 190 proof alcohol, here is your recipe:

Use your measuring cup as follows:

First — Fill your measuring cup to the 1-quart mark and using your funnel, transfer the alcohol into your gallon bottle. Do this three times so that you've transferred three quarts of alcohol to your gallon bottle.

Then — Fill your measuring cup with 19.2 fluid ounces of alcohol. Add water until you reach the 32 ounce (1-quart) mark. Using your funnel transfer this mixture to your gallon bottle. Shake and let it sit for a week.

If you didn't have it before, you now have the alcohol you'll need for this project.

Footnote

Measurements for this project become a lot easier if you switch from U.S. measurements to the metric system. For example, rather than making a *quart* of perfume you could make a *liter*, which is only slightly less than a quart.

My 4-cup measuring cup is graduated in both ounces and milliliters and thus, to produce one liter of 190 proof alcohol, I simply fill it to the 900 ml mark with alcohol and then fill it to the 1,000 ml (1 liter) with water. Now there's no need to estimate where the "29.80 ounce" mark or the "19.2 ounce" mark should fall on your measuring cup.

Better yet, when we get into mixing your oil with alcohol in the next chapter, an inexpensive laboratory measuring column can measure accurately in milliliters and it would be unusual to find any measuring column graduated in ounces.

Distilled water (left); deionized water (right). Either will be fine for this project.

200 or more bottles to the case. In most cases distributors will not sell a single case. Styles you might prefer can be impossible to obtain unless you are willing to spend hundreds of dollars to buy hundreds of bottles when you only want a handful. For this project it makes no sense. What you will need is a vendor who has broken cases down and sells single bottles and bottles by the dozen.

The bottle you will use

Regardless of the capacity of the bottle you choose, it must be one with a threaded neck. Only bottles with a threaded neck can be capped (closed and sealed) with a screw-on spray pump and a screw-on spray pump is the only kind of spray pump that can be attached to a bottle without some sort of capping machine, either manual or automated. Screw-on pumps are easily attached to bottles by hand and, if properly matched to the bottle, will not leak.

Close-up of a threaded bottle neck with a 20/415 finish.

Bottle neck sizes

The necks of perfume bottles are described by their "finish." "Finish" refers to the configuration of the bottle's neck. For threaded neck bottles, finish is given by two numbers. The first refers to the diameter of the neck; the second to neck height and thread style. You will see threaded neck bottles referred to as 13/415, 15/415, 20/415 finish and so on. When shopping for your bottles you will look for a vendor who offers *pumps that match the finish of the bottles you are buying.*

Finish specs have been standardized for bottles but the standards are not always strictly adhered to by manufacturers of spray pumps and caps. It is therefore highly recommended that you purchase your bottles and pumps from the same vendor. Or, if you are buying from two different sources, get samples and *test them* for fit before committing yourself to a large order. You don't want to find — just when you are ready to launch your perfume — that your bottles leak because your pumps are *not* a perfect fit.

Rejected... for now

It's good to know something about the bottles you will *not* be using for this project but which you might use for a future project.

Sprinkler neck bottles

If you were to use sprinkler neck bottles for your perfume you would not need a spray pump. Instead, you could seal it with a simple, inexpensive screw-on cap.

Sprinkler neck bottle. Notice the restricted opening in the neck. I have found these bottles particularly appropriate for men's fragrances.

A sprinkler neck bottle is a threaded neck bottle with an opening in the neck just large enough to splash perfume through it when shaken. A sprinkler neck bottle can be closed with an inexpensive cap and thus can be a money saver. But I have never seen sprinkler neck bottles available in small quantities. The ones I have used required a purchase of several cases. To obtain sprinkler neck bottles it's likely you would confront a $500 or more minimum order size plus shipping charges — and cases of glass perfume bottles are heavy.

For this project sprinkler neck bottles are not a practical option.

Standard threaded neck bottles with orifice reducing plugs

Plastic orifice reducing plug to fit threaded neck bottle.

Orifice reducing plugs are plastic plugs with a small hole in the center. They can be pressed into the neck of a standard threaded neck bottle to make it act like a sprinkler neck bottle. There is the possibility that you could find threaded neck bottles, caps and orifice reducing plugs that go together. But, if you find these three components in a matching size, it's likely you'll have to purchase far more plugs than you need or want, possibly 5,000 or more. Save this idea for a future project when you're producing hundreds of bottles and want to save a bit of money by not using a spray pump.

Threaded neck bottle fitted with orifice reducing plug.

Bottles for crimp-on pumps

Almost all commercial fragrances use crimp-on spray pumps. These are mounted by an automated capping machine after the bottles have been filled. Crimp-on pumps can also be attached to bottles using a manual capping press or other manual device. Some say that a manual capping

Bottles for crimp-on or snap-on pumps. Almost all commercial fragrances will use this style bottle, most commonly with a crimp-on spray pump.

press is far from foolproof and can result in misaligned pumps which mean leaky bottles.

Rather than having a threaded neck, bottles for crimp-on pumps have a slightly extended ring at their neck to which the pump, when crimped, is locked. Once crimped onto the bottle it cannot be removed. These bottles are not refillable.

Bottles for crimp-on pumps can sometimes be obtained in small quantities. It may also be possible to buy crimp-on pumps in small quantities although a commercial case might hold 4,000 pumps and cost a bit over $1,000. If you were producing thousands of bottles of your perfume, it is likely you would purchase a bottle for a crimp-on pump. Pumps and bottles are designated by the size of the bottle's opening, in millimeters.

Bottles for snap-on pumps

Bottles for crimp-on spray pumps can also be used for "crimpless" ("snap-on," or "press-on") spray pumps. A different capping machine is needed to attach them to bottles. Bottle and spray size is designated by the width of the bottle's opening, in millimeters.

Stopper bottles

Historically, fine fragrances were sold in glass bottles with glass stoppers. For the more expensive fragrances these glass stoppers were individually fitted to the bottle making a very good seal. To hold them in place prior to sale, wire, ribbons, and wax might be used, creating a tamper-proof seal. Once opened, the bottle could be resealed with the glass stopper but, while stopper and bottle were matched with precision, the seal was neither air tight nor leak proof.

This arrangement worked well in an age when perfume was used to scent fabric rather than skin. A spray was unnecessary and, in fact, sprays were not used on perfume bottles until the late 1930's and even then they were quite crude. As spray technology and consumer preferences evolved, glass stoppers were mostly abandoned.

Perfume bottles with glass stoppers. On the left is a modern bottle with its glass stopper fitted with a plastic plug. On the right is an older bottle with a glass stopper ground to fit the bottle.

Today bottles with glass stoppers are sometimes seen (I have some) and they do have a very elegant look. They can be purchased in small quantities and the stoppers are generally fitted with a plastic gasket to make a tight seal as the bottles and stoppers are mass produced. A stopper dedicated to it's own particular bottle is a thing of the past.

Stopper bottles look nice but, if you were to try them, you would find that this is not a practical choice.

Chapter 7
Fine mist spray pumps

The only fine mist spray pumps that will be practical for you, for this project, are those that will screw onto a bottle with a threaded neck and matching finish (i.e., neck configuration). Your challenge will be to find threaded neck bottles and screw-on pumps that will fit them. If you can, get both bottles and pumps from the same vendor. This is the surest way to avoid the small misfits that can cause leaks. Sometimes you may find what you want from two different vendors but in this case be sure to get samples and test for leaks, before you use this combination to bottle any perfume you are going to sell. You don't want to discover a problem *after* you have started taking orders.

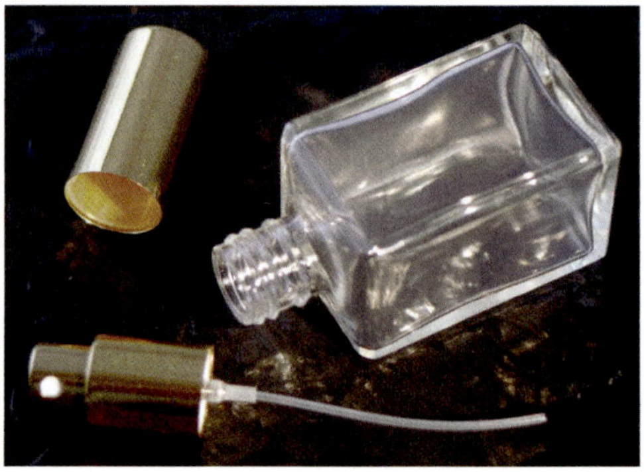

Threaded neck perfume bottle with screw-on spray pump and overshell (also called a "cap"). The bottle shown has a 2 fluid ounce capacity and both pump and bottle have a 20/415 finish.

Alternatives

Caps

A screw-on cap is a low cost alternative to a more expensive spray pump. A cap, however, can only be used with a sprinkler neck bottle or a threaded neck bottle fitted with an orifice reducing plug. In either case the cap must be the correct size to fit the bottle's finish.

Bottles for crimpless and crimp-on pumps

Crimpless and crimp-on pumps require a bottle with a non-threaded neck which instead has a ring of glass to which the pump is attached. These necks are commonly described by a single number, the size of the opening in the bottle's neck measured in millimeters.

Bottles for crimp-on and snap-on spray pumps have necks which are not threaded but instead have a ring (which is called a thread!) to which the pump will be attached.

Crimp-on pumps

Crimp-on fine mist spray pumps are the industry standard and major filling houses without exception will have automated equipment that will fill and cap these bottles with crimp-on pumps as they pass down an assembly line.

Crimp-on pumps can also be attached to bottles with a manual capper although some issues have been reported with these tools such as deformed pumps, which cause (filled) bottle and pump to be junked as, once mounted even badly, the crimp-on pump cannot be detached from the bottle.

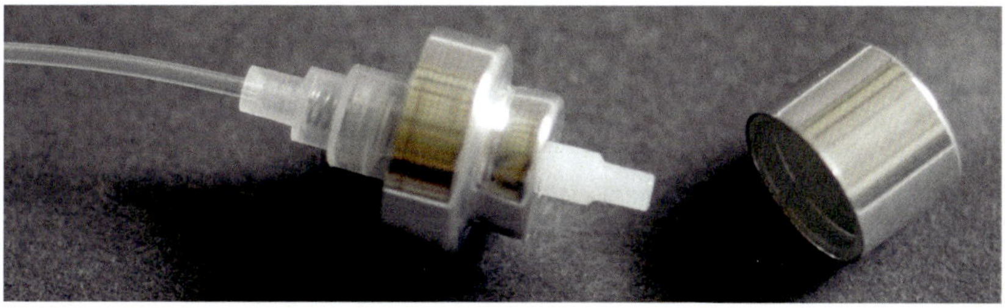

Crimp-on spray pump. The actuator (the part that gets pressed to release the fragrance) is shown at the right. The large gold band around the body of the pump is the ferrule, the ring that gets crimped around the bottle. The dip tube, which picks up the perfume from the bottle, extends downward from the pump into the bottle.

Crimp-on pumps have a lower profile than screw-on pumps which can give the finished bottles a more elegant look. All of the bottles you see displayed at major perfume counters are fitted with crimp-on pumps.

Crimpless pumps

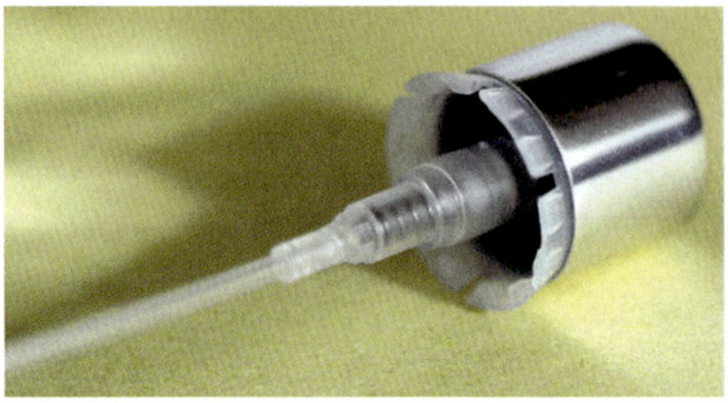

Snap-on (crimpless) spray pump showing "claws" which will hold it firmly to the bottle and dip tube which will extend into the bottle to draw up the fragrance.

Crimpless pumps — also referred to as *snap-on* or *press-on* — are a temptation as they can appear deceptively easy to press onto a bottle. I've tried to do it by hand and by hammer with poor results. Unless you have a correct press for mounting these pumps you risk breaking bottles and smashing the pumps themselves.

Crimpless pumps have a lower profile than either screw-on or crimp-on pumps. They give your bottle a very nice look.

Footnote

Several sources now offer relatively affordable manual capping tools that are appropriate for small production runs. In the months ahead it may become practical for small producers to make use of crimp-on and crimpless pumps.

Chapter 8
Filling and capping

You have your perfume; you have your bottles and pumps. Now it's time to fill the bottles and attach the pumps. For the pumps it is simply a matter of screwing them onto the bottles. As to filling the bottles with perfume, you can do it any way that works for you but I'll share with you the technique I was introduced to some years ago and which I still use. If you follow these steps I think you'll understand why I continue to do it this way.

Finished perfume drawn from measuring cup with a metal tipped turkey baster. The sixteen bottles shown hold two fluid ounces each. The baster picks up about one fluid ounce of perfume with each squeeze.

Using your 4-cup glass measuring cup, fill it about 2/3rds full with your perfume. Set the bottles you are going to fill on a paper towel on a desktop or table that won't be damaged by spills of perfume. Using your turkey baster, fill it with perfume from your measuring cup. The baster will hold about an ounce of perfume.

Insert the tip of the turkey baster into the neck of each bottle and squeeze the bulb gently releasing the perfume. Try to do this in a smooth motion. Don't let the tip of the baster rest on the bottom of the bottle. The idea is to get a smooth flow of perfume and avoid splashes which cause waste.

To fill each bottle, fit the tip of the turkey baster into the neck of the bottle and squeeze the bulb.

Fill each bottle up to its shoulder but not beyond. It is essential that you leave air space between the perfume and the pump. If you overfill the bottle and fail to leave adequate air space, on a hot day the perfume will expand and force its way past the pump. This will cause leakage and, under some circumstances, will blow the pump off the bottle. Be warned!

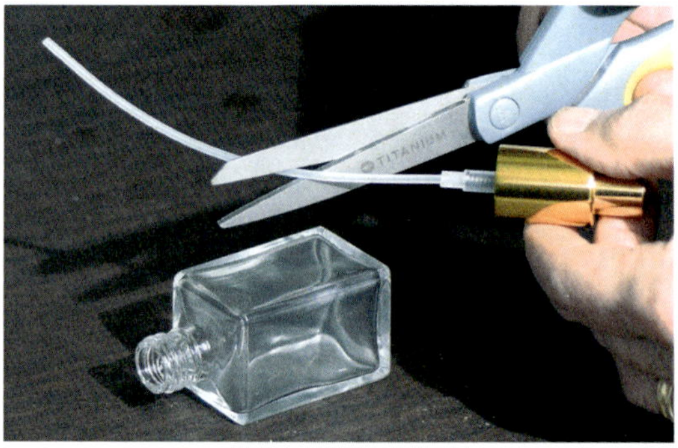

Using scissors, cut the plastic dip tube to reach the bottom of the bottle and then a small bit extra. Cut the dip tube at an angle so that it won't become plugged if it touches the glass inside the bottle.

Perfume bottle filled to the shoulder.

One more thing. Before screwing your spray pump onto your bottle, trim the dip tube to an appropriate length. You can cut it with a sharp pair of scissors. Leave it long enough to reach to the bottom of the bottle and then just a bit more. Cut the dip tube *on an angle* so that, should it press against the bottom of the bottle, it will still be able to suck up — and spray — your fragrance.

Once the bottles have been filled and the pumps screwed on, I like to wipe each bottle with a paper towel dampened with alcohol to remove any perfume that might have leaked onto the outside of the bottle during the filling process. You want your bottles to be clean and dry before you attach labels.

Once the bottles are filled and wiped, let them sit for an hour or so before you attach your stickers. This will allow any remaining alcohol on the bottle to evaporate.

Chapter 9
Labels

Once your perfume is bottled you'll want to label your bottles. Three issues arise. First, what will you *call* your perfume? Second, what *text* will you put on your labels? Third, how will you *produce* your labels?

Naming your perfume

If you only plan to distribute your perfume to friends and family and have no plans to advertise it on social media or elsewhere you can pretty much give your perfume any name you want as you won't be stepping on anyone's toes. If you plan to promote and *sell* your perfume the story is quite different.

When you slap a name on your perfume bottles and offer them for sale, the name becomes a trademark and that trademark becomes your property UNLESS that name or a similar name is already in use by someone else. In this case you could be infringing on *their* rights and, if caught, you will have a problem. Thanks to the power of the internet, if you use a name for your perfume that somebody else is already using there is a good chance you will be caught, even if you think you're too small for anyone to notice.

As an example of the reach of the internet, I was marketing one of my fragrances on a very small scale and was contacted, by phone, by someone with an interest in the name. The name was rightfully mine but the other party wanted to use it and agreed to buy my rights to it. Their interest was a big surprise to me but I was happy to take the money. The simple lesson is that once you go online, anywhere, with your perfume's name you are not anonymous! Be warned.

When you put your perfume up for sale with a name not yet taken by anyone else, you establish a legal right to that name which is a trademark. You do not have to register the name — the trademark — to obtain this legal right. If you are investing a good deal of money in your project and anticipate large sales, you might want to register your trademark and it is advisable that you do so with the assistance of a lawyer specialized in trademark law. For this project trademark registration and the required follow-up filings will generally be an unnecessary expense and bother.

Checking for prior use

How do you know if anyone else is using the name you want to use? I research it two different ways. First, I do a Google search for "[my name] perfume." If nothing turns up I continue with search terms such as "[my name] cologne" and "[my name] cosmetics." When you search, pretend you are really trying to find a product with the name you want to use. If nothing turns up, this is a good sign.

From Google I go to the U.S. Patent and Trademark Office website and to their TESS (Trademark Electronic Search System) database where again I search the name I want for my perfume. (TESS is free for anyone to use and you don't have to register.) If nothing comes up I'm pretty confident my name will be good. If something does come up I look to see what class is represented by the trademark. If it is not "IC 003" (International Class 003 includes perfume) I am still okay. At this point I'll consider the name to be free for my use.

Everything else on your label

Aside from the name of your perfume you'll need a bit more text on your label to comply with FDA (U.S. Food and Drug Administration) requirements. All is quite straight forward and should cause you no problems. Outside the U.S. fragrances have different labeling requirements.

First you'll want to indicate the size of your bottle (which will be the volume of the contents) in fluid ounces or milliliters. This will be no problem as you picked the size when you selected your bottles.

Label showing fragrance name, description (...an original *fragrance...*"), and other required information.

Then you must show your company and address. If you don't have a company you can use your own name. For address you can just show city, state and zip code if that will be enough to someone to find you. If not, include your street address. Then there's a line to show your relationship with the contents. If you were just distributing a fragrance by someone else (which is not the case here) you would add the line "Distributed by [your name]." As this is *your* perfume and *you* are producing it, you will use a line such as "A perfume by [your name]" or just the word "perfume" or "fragrance." Finally you'll list the ingredients of your perfume. In the United States (as of 2021) this continues to be simple: "Ingredients: SD Alcohol, Fragrance"). The alcohol is listed first because it supplies the larger portion of the contents.

Printing your labels

There are commercial printers who are more than happy to print your labels once you have prepared the artwork. Some will even prepare the artwork for you. Usually this will mean printing at least 1,000 labels and you might find it difficult to match the standard sizes offered by printers with the area to be labeled on your bottles. You may need a custom size and this will cost more money. If you want your label printed in more than a single color, this will cost more money.

For a small number of bottles — anything under 500 — I generally print my own labels on sticker paper using an inkjet printer. This allows

Chapter 11
Wrapping it up

If you've been reading the chapters in order but haven't yet carried out any of the instructions, it's time for you to get busy. You now know all you need to know to put a perfume together — a perfume you can sell... commercially... or just enjoy with your friends.

If you were making 10,000 bottles of perfume you would be following these same steps. The difference would be that outside contractors would be doing much of the work you are now doing yourself, and they would be using automated equipment instead of the crude (but very effective) equipment used here.

In making your first quart of perfume you may find yourself purchasing more supplies than you need for this project. That's fine because you can use them for your next project. These "extra" supplies have been particularly useful for me as I develop a number of fragrance oils with the intent of producing only a dozen or so bottles of each. With alcohol, water, bottles, and pumps on hand, my only expense for these new fragrances is the cost of the essential oils and other aroma materials I use.

If you're still a bit hesitant about making your own perfume, I urge you to set aside your fears and unleash your creative muse. You may discover skills you never dreamed were yours and the enormous satisfaction of producing a great perfume!

Appendix A
Supplies

All materials used in this project were acquired through the internet from U.S. sources or sources that ship to the U.S. Listed below are the source used for this (July/August 2021) project.

Vendors can come and go and some I've used in the past are no longer in business. Others are in business but no longer accept small orders. For current vendor information visit my website:

www.PerfumeProjects.com

and follow the link to the "Vendors" section.

Sources of the supplies and materials used in this project:

SKS Bottle and Packaging, Inc.

https://www.sks-bottle.com/

I've used their Amber Glass Boston Round bottles in pint, quart, and gallon sizes. The pints when making new fragrance oils, the quarts for making small quantities of perfume (as in our current project), and the gallons when I like a fragrance enough to make more of it. Here are SKS's offerings:

Pint bottles, packed 12 to a case, caps included.

Quart bottles, packed 12 to a case, caps included.

Gallon bottles, packed 6 to a case, caps included.

Scentsational Shop

https://www.perfumeoils.com/

Perfume bottles (threaded neck).

Screw-on spray pumps.

Bottle/pump combos.

Save On Scents
http://www.saveonscents.com/

Alcohol, SDA 40B.

Fragrance oils (for this project.)

Fragrance oils (sample packs.)

CQ Concepts
https://cqconcepts.com/

Deionized water.

From your local supermarket or gourmet shop

4-cup glass measuring cup graduated in fluid ounces and milliliters.

Funnel, glass or metal.

Turkey baster with metal tip.

Coffee filters.

Distilled water.

Other useful sources

McKernan Packaging Clearing House

https://www.mckernan.com/

Bottles and closure closeouts and bulk sales. The sprinkler neck bottle shown on page 24 came from McKernan.

Plush Folly

https://plushfolly.com/

A UK source of aroma materials and perfume bottles including bottles with snap-on sprays.

Perfumer's Apprentice

https://shop.perfumersapprentice.com/

Aroma materials, bottles, and closures.

PerfumersWorld

https://www.perfumersworld.com/

Aroma materials, software, instruction and equipment.

Made in United States
Orlando, FL
07 November 2026

90337540R00031